Order ID: 111-4834465-0617006

Thank you for buying from Eastons Books Inc on Amazon Marketplace.

Shipping Address:	Order Date:	Thu, Sep 24, 2026
AMZ ORDERS APTP	Shipping Service:	Standard
800 AVONDALE AVE	Buyer Name:	Rhenoa
COLUMBUS, OH 43212-3473	Seller Name:	Eastons Books Inc

Quantity	Product Details
1	**Field Effect [Paperback] [1992] unknown author** **SKU:** 37825 **ASIN:** 0943975026 **Condition:** Used - Very Good **Order Item ID:** 170727957307161 **Condition note:** Paperback in Very Good+ condition.. Maggie Valley, NC. The Centric Foundation, Inc.. 1992. 64 Pages. Includes 12 page Precis of Field Effect: the Pi Phase of Physics. Manuscript of the full equation.. * Quick Shipping * All Books Mailed in Boxes * Free Tracking Provided *

Returning your item:

Go to "Your Account" on Amazon.com, click "Your Orders" and then click the "seller profile" link for this order to get information about the return and refund policies that apply.

Visit https://www.amazon.com/returns to print a return shipping label. Please have your order ID

$2.50

Precis
of
Field Effect
The Pi Phase of Physics

By Leigh Richmond Donahue

Manuscript
of the
Full Equation

FIELD EFFECT, the PI Phase of Physics
Published by
Dimensional Sciences, Inc
PO Box 167
Lakemont, GA 30552
ISBN #0-943975-6
$9.95

The Centric Foundation, Inc

Centric Analysis: The mathematics of Trigonometry applied to field theory and relativity

Leigh Richmond, Pres.
R.V. Donahue, Exec. V.P.
Scott Richmond, Sec/Treas
Tucker L. Stoodley, V.P.

P.O. Box 9
Maggie Valley, NC 287
(704)627-98

Precis of
FIELD EFFECT, The Pi Phase of Physics

by Leigh Richmond Donahue

(FIELD EFFECT is published by Dimensional Sciences, Inc.
PO Box 167, Lakemont, GA 30552
ISBN# 0-943975-6; $9.95)

It is **not** $E=MC^2$. It is $M=EC^2$.

It is electomagnetic energy, moving at the square of the velocity of light in a tight, dense ring at the core of the electron that is equivalent to matter; that is, in fact, the matter that makes up the galaxy.

Our galaxy, ourselves, and the electrons of which we are built, function on a logical pattern and operative system.

That pattern and that operative system make common sense and can be understood.

Once you find that pattern, the rest falls into place.

There are not anomalies.

Th pattern is built of two factors: the structure of the electron, and the galactic system for the storage and usage of information, as we shall see.

The electron is structured of a core, which is matter, and the field which surrounds it.

It is the field that surrounds the core that makes possible the cohesiveness of our galaxy and our selves in a dynamically stable form, and that makes possible as well the storage and exchange of information necessary to evolution.

We will start our analysis with the system by which the electron came into being; then its structure.

That will be followed by an analysis of the pattern of information storage and usage, which made possible evolution.

COMPONENTS OF THE ELECTRON

The electron is built from four components:

1. The ether: A sea of energy without motion which exists throughout the universe.

2. Energy in motion: What we call an electrical flow. A twisting thread of energy moving through the sea of ether.

3. Magnetic force: Ether forced into motion by the electrical flow. This takes two forms:

(a) As the electrical thread pushes aside the ether, it

pulls some of the ether into motion around itself in a spiral coil of magnetic force, along its length, much like a coil of wire. This coil is at right angles to and moves in the opposite direction to its electrical center. As the energy of the electrical flow increases, the coil tightens. When the energy relaxes, the coil loosens. This is known as frequency variation, and can be seen in a comparison between radar and radio wave.

(b) Meantime the magnetic coil causes ripples in the sea of ether around the now electro-magnetic flow. These ripples spread out around the electromagnetic path, decaying towards infinity, like ripples from a stone dropped into a pond. This is the magnetic field.

4. Vacuum. Where electromagnetic flows have cleared areas in the sea of ether.

None of these four factors is particulate. In other words, they are not made up of particles... not the sea of energy without motion (the magnetic shield and its magnetic field) nor the vacuum created in the ether by the various flows.

Granted these four components, the structure of our alternating current universe becomes understandable.

THE PATTERN

To build a galaxy, you need a stable particle as its base.

The pattern of that particle must be obvious in everything from the structure of the elements to the galaxy, including the various solar systems, the rocks and humans

Such a particle must be of a configuration that cannot easily be destroyed. It must have a gyroscopic configuration and precess when confronted with obstacles, rather than battering or being battered. It must create itself continuously from the forces at hand.

It must be in motion, thus qualifying as energetic. It must have a push and pull of its own (polarity). It must respond to both a push (a kick in the pants) and the pull of a vacuum or of surrounding energies. It must have a near-infinite lifetime.

Such is the electron.

There are an infinitude of minute particles, with half-lifes in millionths and billionths of a second -- quarks, charms, bits and pieces. They do not meet the criteria. They are strays.

It is the electron that is stable, and it is the pattern of the electron that can be found through the periodic table of the elements; through the molecules; up through the rocks, the animals and you. It is the pattern of the galaxy.

The electron is constantly being created in its infinite numbers, using the four factors that are at hand. They are created by the frictions at the center of the galaxy, and thrown off into the long arms of the flat spiral whirl that makes up our Milky Way.

THE ELECTRON

Take the twisting thread of electrical energy moving through the ether, moving the ether aside, which creates the magnetic coil that surrounds its length. It builds, as well, a bow-wave of ether, which condenses into a wall. Meeting that wall, the thread of energy turns in a circle to meet and pass its own tail. It is still surrounded by its coil of magnetic shielding at right angles to itself and moving in the opposite direction, which keeps the electric threads from combining.

These threads of now electromagnetic energy form a ring, their centers flowing in one direction, their magnetic shields flowing in the other. The coils of the outer magnetic "wire" mesh, pulling the electromagnetic flow into a tighter and tighter ring.

This is the snake of Chinese antiquity that swallows its own tail.

As it tightens, the ring becomes dense and denser, until it creates so strong a vacuum at its center that an amount of ether equal to the amount of energy in the ring, is pulled in at one side, through the center, and out the other side.

With this opposed flow going through its center, the ring cannot either become smaller or collapse. This small, tight, extremely dense ring of electromagnetic force, moving at the square of the velocity of light, is the core of the forming electron.

Now we look at the opposed flow, pulled through the center of the core from the surrounding ether. As this opposed flow of ether goes into motion, its threads each form a spiral coil of magnetic shield.

This opposed flow, this dialectromagnetic (opposed electromagnetic) flow spirals through the center of the core, jets out spherically to a tremendous distance (in respect to the core) and is walled back as it meets the wall of surrounding ether, to the vacuum at the nose of the core-ring, and through again in a raging, pico-sized torrent. It flows at several orders of magnitude less than the velocity of the core, and is voluminous rather than elements; through the molecules; up through the rocks, the animals and you. It is the pattern of the galaxy.

The electron is constantly being created in its infinite numbers, using the four factors that are at hand. They are created by the frictions at the center of the galaxy, and thrown off into the long arms of the flat spiral whirl that makes up our Milky Way.

THE ELECTRON

Take the twisting thread of electrical energy moving through the ether, moving the ether aside, which creates the magnetic coil that surrounds its length. It builds, as well, a bow-wave of ether, which condenses into a wall. Meeting that wall, the thread of energy turns in a circle to meet and pass its own tail. It is still surrounded by its coil of magnetic shielding at right angles to itself and moving in the opposite direction, which keeps the electric threads from combining.

These threads of now electromagnetic energy form a ring, their centers flowing in one direction, their magnetic shields flowing in the other. The coils of the outer magnetic "wire" mesh, pulling the electromagnetic flow into a tighter and tighter ring.

This is the snake of Chinese antiquity that swallows its own tail.

As it tightens, the ring becomes dense and denser, until it creates so strong a vacuum at its center that an amount of ether equal to the amount of energy in the ring, is pulled in at one side, through the center, and out the other side.

With this opposed flow going through its center, the ring cannot either become smaller or collapse. This small, tight, extremely dense ring of electromagnetic force, moving at the square of the velocity of light, is the core of the forming electron.

Now we look at the opposed flow, pulled through the center of the core from the surrounding ether. As this opposed flow of ether goes into motion, its threads each form a spiral coil of magnetic shield.

This opposed flow, this dialectromagnetic (opposed electromagnetic) flow spirals through the center of the core, jets out spherically to a tremendous distance (in respect to the core) and is walled back as it meets the wall of surrounding ether, to the vacuum at the nose of the core-ring, and through again in a raging, pico-sized torrent. It flows at several orders of magnitude less than the velocity of the core, and is voluminous rather than around and back in at the nose.

A's field, spiraling in its counter-rotation to B's field goes around B's core and between the two cores, cushioning them apart. It continues its spiral through its own core, out and around within B's field, thus creating a shell within a shell as it spirals back into the nose of its original core.

The energies are violent, their structures, motions and velocities opposed. The two electrons are hopelessly interlocked.

This is the proton.

What magnificent and complex furies at sub-microscopic levels and in micro time are to be found in the elements that go to make up our beings.

THE NEUTRON

What happens when electron meets electron nose-to-nose? In other words, black hole to black hole.

If their handedness is the same -- if both have either left-hand or right-hand spiraling fields, they swirl away from each other. The winds that create the suctions of their fields tip them apart and throw them in opposite directions.

But when a left-handed electron meets a right-handed electron, nose to nose, they dance in tune.

The violence of the field entering each of the black holes cushions them apart. The suction of the black holes keeps them together. The jet of each pushes it against the other.

This is the neutron.

THE ATOM

It is now possible to structure the entire periodic table of the elements.

We begin with the hydrogen atom, and the proton from which it is formed -- the simplest proton, with only one shell, which lies between the field of the original electron (A) and the field of the captured electron (b).

Within this shell and the violence of its conflicting currents lie all the ingredients for the creation of an electron. It is inevitable that a new electron will be formed. The core of this new electron is dense, and will swing down to orbit the equator of the proton.

It is said that the electron has 1/1837 the mass of the hydrogen nucleus; but that its volume is 1837 times a great.

So now we must look at a definition of mass, as versus that

of the field.

Mass is the dense core of the electron, and/or the core of the proton.

It is the core that is dense to near solidity.

The field(s) are, to all intents and purposes, invisible. There are so light and voluminous in nature that they are measured as a charge.

We are speaking of micro-time and micro-space, and their velocities.

The proton core, or nucleus, remains at the center of the atom, measurable as itself.

The electron core, orbiting the nucleus at tremendous velocities, has the apparency of a circular volume, the volume of the entire orbit -- 1837 times as great as that of the central nucleus.

The hydrogen's orbiting electron swings on its field as on wings, that stretch from pole to pole of the atom -- but always within the shell to which it is confined. Those wings are of violent motion, but so voluminous and so light a density that they register only as a charge.

Thus the hydrogen atom, the start of the ladder of the periodic table.

To begin the atomic climb up the periodic ladder, it is necessary to add energy to the hydrogen nucleus.

A proton adds energy and mass in the same manner by which it originally became a proton. It captures another electron from outside, through its black holes, into a central core.

The core of the newly captured electron is stuck in the throat of the proton, and its field forms a new, counter-rotating shell around the earlier shell(s).

With this added energy and mass, the spin of the proton increases, its becomes internally denser and smaller; but the new shell brings its volume back to the original size.

Within this new shell, again, the violence of conflicting, counter-rotating currents creates new electrons, their cores swinging in to orbit the equator of the nucleus, while their fields expand to fill the shell.

Where there are multiple electrons within a shell, they may combine to become neutrons, positrons, etc., as well as remaining electrons; but their number is restricted to spatial characteristics within the shell to which they are confined, since their expanding fields will fill the space available, and obviate the entrance of new electrons.

Thus you have the means and structural capabilities of the ladder which we know best as the Periodic table of the elements... the atoms being built through time and evolution from the ether, through the basic stable particle, the electron to the simple hydrogen atom -- building blocks of the universe.

WEBBING INTO FORM

The core of each electron is isolated within its field. The cores of the atom are each isolated within its own field, as well as within the field of the atom itself.

The outer shielding field of each atom, composed of the twisting "superstrings of dialectro-,agnetic force interacts with the outer superstrings of other atoms. Where these fields intersect, they weld, tying the two atoms together.

The intersecting fields group the atoms into the overall network of intersecting fields that are forms -- molecules, chemicals, dust, rocks, crystals, groupings; then planets, solar systems, and eventually on up to the galactic spiral, a long distance network of more tenuous interlinks.

Around the various forms netted together into a unit, an overall field develops, which in turn nets into the larger and larger networks.

thus the cores of each object remain isolated, while the superstrings of their fields net them, step by step, into the larger and larger networks of surrounding forms.

EVOLUTION 101

A computer is a tool for receiving, comparing, relating, integrating, storing and demonstrating information; and for holding it ready to be accessed.

There are two forms of computer -- digital and analog.

A digital computer handles information in exact quantities and in static relationships. Its facts can be refined to the umpteenth decimal point.

An analog computer stores information in variables and in time relationships. It handles time not as sequences, but as an amalgam. It deals in fluid relationships, such as the relationships expressed in Pi.

Neither form of computer is sentient or self-aware.

The electron can be shown to be structured of the two forms of computer, digital and analog, interlocked but not interacting.

The core of the electron has the pattern of a primitive, unprogrammed digital computer. It is formed of electronic strings coiled in a tight, high velocity ring. Each electronic string is in a sheath of magnetic flow which moves (at the same high velocity) in the opposite direction. Its polarities are external, and it responds to external electromagnetic frequencies. It does not respond to the torrent of dialectromagnetic force pouring through its center. That force is invisible to it.

This core of the electron, this primitive digital computer, is similar in submicroscopic form to today's integrated circuit digital computers.

The field of the electron is a primitive, unprogrammed form analog computer. It is structured of dialectromagnetic "superstrings" -- i.e., strings of scalar wave frequencies which are internally balanced and sero-aum, so that they present no external hooks. These superstrings will interact only with other scalar wave frequencies, and at a point where they intersect and create a focus. The electromagnetic core through which these scalar wave frequencies pour before they spread into a sphere forming a gyroscopic particle, is invisible to the superstrings.

This field of the electron, this primitive analog computer, is similar in pattern to today's developing photonic computers with their lasers that do not "see" matter. It stores its information in holographic form

The two computers -- digital and analog -- that make up the gyroscopic electron, are opposed in direction of flow, in position, in volume, in density and in velocity. The tight, dense ring of force that is the core circles at a velocity several orders of magnitude greater than that of its voluminous, diaphanous englobing sphere. The two, though interlocked, are effectively invisible, the one to the other.

The two computers are also opposed in function. The core measures the input information from external frequencies in exact quantities and in static relationships. Digitally. The field registers information input from its focal contacts (the outer fields of other electrons, atoms and other forms on up the ladder) in fluid relationships over frames of time, in the variables of Pi, and in holographic form.

These electrons, each formed of the two opposed computers and in gyroscopic motion, combine to form protons, atoms and elements up the periodic table. Each electron retains its own form within the field of the atom which it helps to build and of which is becomes part, so that the atom, too, has the form of interlocked digital and analog computers. The atoms in turn retain their forms as they build into molecules, which also in turn have overall fields containing the atoms and are in the form of interlocked digital/analog computers. And so on, up the evolutionary ladder.

Computers within computers within computers in a vast variety of interwebbed forms.

Here at the base of the evolutionary ladder you have, illustrated, the pattern and operative system on which our galaxy is built.

The cores of each of the atoms, molecules, and larger forms are isolated, the one from the other.

The outer fields of each component intersect, find total points and weld. If the two intersecting superstrings are of fairly well-balanced components, they remain welded. If one of the components is sufficiently heavier than the other, the weld breaks.

Take, for instance, the atoms of a crystal. Here you have a myriad of atoms webbed into a whole, each atom in violent motion, the myriad cores at (comparatively) planetary distances the one from the other, yet the crystal is in a dynamic stability, webbed into a unit by the intersecting focal points of its fields. Around this crystal unit, its own field develops to be its outer shield.

So, too, for the galaxy, which is formed with the same pattern and the same operative system.

At the galactic center, as the galactic field pours through the galactic core, the friction between the two forces creates electrons that are thrown off into long spiral arms extending out from the core within the swirling galactic field.

Over the billenia, those electrons, forced outward, are webbed into suns and planets. As they reach towards the ends of the spiral arms, the netted forms evolve towards life.

Thus has begun the long, slow climb over billennia from electron to form -- to fire, air, earth and water.

If you thing that it's all quiet on the galactic front while this is happening, you're very much mistaken.

There is electronic fire, collisions, the atoms have formed into chemicals and their fires; the pull and push of the vacuums, the forming and destruction as the galactic form grows.

And always and always, the torrential winds of the galactic field forcing its way through the black hole at the center of the galactic core, is throwing off new electrons to spiral out and away in long spiral arms.

The core(s) of the galaxy itself we see as the Milky Way. But the galactic field -- the huge, webbed field of superstrings that englobes those cores is invisible to our eyes.

The electrons whirl away from the galactic core at the velocity of light, the particles careening through space, driven by the jets at their tails, tossed and turned by the violent galactic winds, comparable to our own system's solar winds.

Further along the spiral arms, where suns and then planets have formed, they slow to a more ponderous velocity.

There is electronic fire and chemical fire, and terrible volcanos as magnetic flux within the crusts of the suns and planets throws out melted iron and rock. There is chaos, but form has begun.

By the time (billennia) that the galaxy is fairly well formed, its webbing of fields -- in what I call the "net" -- not only holds the vast variety of forms together, but webs the galaxy into a whole, the stars in their courses, the planets that surround them, the long arms of the spirals -- all within the overall galactic field.

Each net holds the holographic record of its formation and its history, and is interconnected with the nets of every other particle, in a spiderweb of information available, ready for access.

Each core, isolated, holds the electromagnetically stored information of its own formation and history.

This galaxy of information is interlocked, each field to the network of fields, each field to its own core.

But as yet there is not life, no interaction between fields and cores; no sharing of information between the digital and analog storage libraries.

There is duplication of form, since the attractive forces of the various forms tend to duplication; but there is no reproduction.

The computers within computers within computers are interlocked, but not interactive.

Life begins when a means develops for the sharing of the two different types of knowledge. It begins when exact pieces of static information are combined with the variables of motion and continuation through time.

It begins when an interface forms between the two creates the combination of exactness of form with variable motion through time, as well as a means of preserving, down the centuries, the conclusions that prove themselves valid.

Take the amoeba. A primitive interface develops between

the overall field of the amoeba, and the body at its core.

Each computer presents its information at the interface. When the two sets of information corroborate -- reinforce -- each other, the results are returned to both computers and stored, each in its own system of handling. Where there is a discrepancy -- where the two do not reinforce -- each set of information is sent back to the originating computer and stored as unreinforced, probably negligible.

Once this communication is established, a life form becomes possible. Reproduction becomes possible. It becomes possible to vary the relationships of exact components, to create new patterns in individual motion through time and space, preserving the viable, discarding the unviable.

When the amoeba splits to reproduce, it splits along the interface, taking the duplicated information that has been corroborated with each half, as well as half of each of the two computers.

At this point, a support structure for each of the two computers begins to develop -- the body for the digital computer, the overall field for the analog computer.

As life evolves, the intimacy of interaction at the interface between the two computers develops slowly into sentience, self-awareness and eventually intelligence.

The intimacy of interaction between the two computers is an exact measure of the intelligence of a life form.

In the human, the core, or digital computer, is stored in the skull. The body acts as a support structure to that computer.

In the human, the field, or holographic computer is stored in the aura, which surrounds and interpenetrates the skull.

The support structure for this computer is the field which surrounds and interpenetrates the body, with the chakras as points of contact; and with the superstrings of the field interwebbed into the net of superstrings that connects the field to its environment, and to the vast network of fields throughout the planet, the solar system and the galaxy.

In the human, the interface between the core computer and the field computer is thought possibly to be the pineal gland, a silicone structure which shatters at death of the body.

The body, or core, of the human, is isolated within itself except during the sex act, when it combines and exchanges information with another human. The body continues its own specialized form and intelligence by reproducing its own kind with the delicate network of genes to give it immortality.

The immortality of the field is achieved at the death of the body, when the field forms a hologram of a body through which it can continue an interlock and interaction. It goes out on the net until such time as it finds another body with which to amalgamate, whether in the past, the present or the future.

The two forms of intelligence -- digital and analog -- are contrasted in scope, the one isolated within its field, the other capable of reaching out across whatever distances and times it can encompass, to gather information much as you would with a telephone system or fax.

They are contrasted in frequency. They are out of phase, the one with the other. The body (we are speaking now of the human) has recapitulated the evolutionary sequence at the time of its birth, and is ready to begin maturing. The field begins recapitulating its evolutionary sequence at the birth of the body.

As the body reaches maturity, the field reaches the ability to begin maturation. As the body passes the reproductive stage, the field reaches maturity. As the body dies, the field has become mature and is ready to go out on the net as a holographic form.

We have been speaking of the body, but let's return to the vast form of the galaxy, also with the two computers, quite similar in pattern to the human, creating and recreating itself constantly as it throws off electrons at its core into the long spiral arms that flow outwards within the overall structure of the its tremendous field component.

Are the two vast computers of which our galaxy is formed merely interlocked? Or have they created an interface and developed the intelligence that such an interface makes possible?

Has the galaxy itself become a Supreme Intelligence of which we humans and the other intelligences that must exist towards the ends of the galactic spiral arms a sub-microscopic part?

Or is the galaxy itself simply a tremendous informational source, with the humans and other life forms acting as a primitive galactic interface?It's quite a question. The computers are there, and can be shown to be there. The interface potential is there, and can be shown to be there.

And we -- and the other intelligences at the far ends of the spiral arms -- the interface which makes the information useful? Or has the galaxy itself, as such, the interface and intelligence capability?

FIELD EFFECT

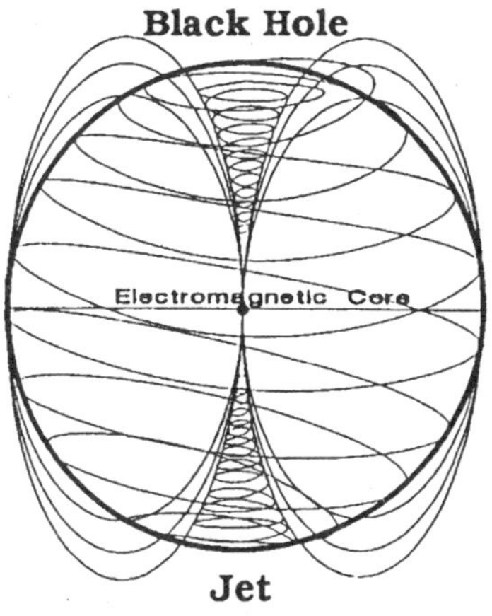

The Electron, the earth, the galaxy, and all other living things. Whether you're referring to an Angel or an Apple a Rock or a Frog; The "Field" is complete...around and through each and every one.

Dialectromagnetic Field

FIELD EFFECT
Proton – Interior Configuration

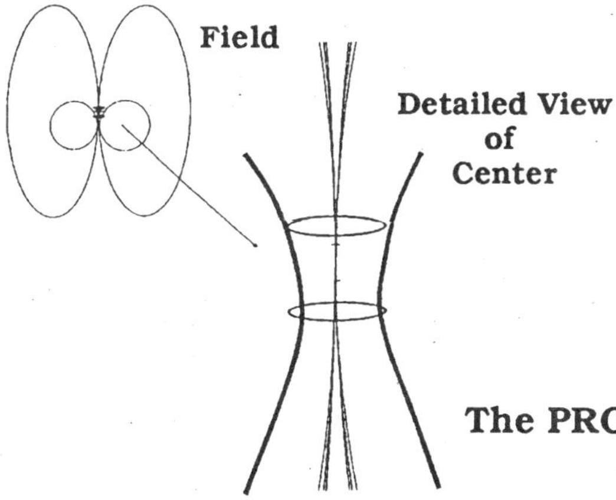

The PROTON

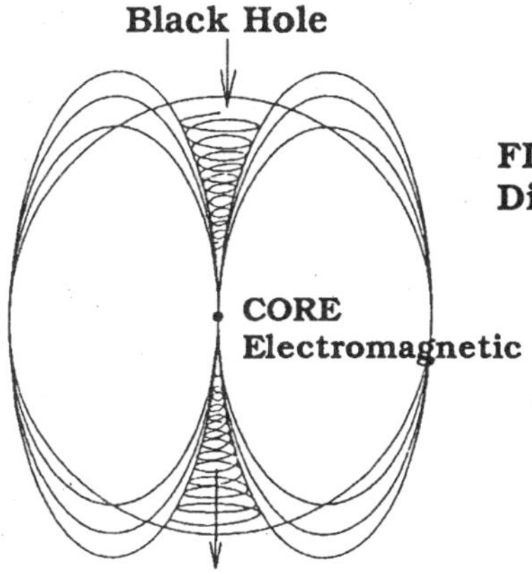

FIELD
Dialectromagnetic

The electromagnetic core and the dialectromagnetic field – equals and opposites, interlocked – operate as a unit of opposed forces.

The ELECTRON – Shown in two dimensions

FIELD EFFECT

. . . The Pi Phase of Physics

by Leigh Richmond Donahue

The Centric Foundation, Inc.
P. O. Box 908
Maggie Valley, NC 28751
(704) 627-9822

Field Effect - the Pi Phase of Physics

Published by
Dimensional Sciences, Inc.
P O Box 167
Lakemont, Georgia 30552

ISBN # 0-943975-02-6

Printed in the United States of America by
Currahee Printing Co.
Toccoa, Georgia

Books co-authored with the late Walt Richmond:

The Lost Millennium (Published also as Siva)
The Probability Corner
Gallagher's Glacier
Challenge the Hellmaker
Phase Two
Positive Charge
Phoenix Ship

Space and time—the electromagnetic and the dialectromagnetic—are the great opposed polarities. The opposed polarities of charged particles are the resultant of this dichotomy.

From: Findings in Centric Analysis

FIELD EFFECT

Dedication

To Nancy Lodge and
Jean Urban –
who made it possible;
To the late Walter F. Richmond,
basic research physicist,
who taught me the esoterics
of electronics and physics;
To Peter Kelly, genius of radionics,
who taught me the usages of the
dielectro-diamagnetic spectra.
And to
Cdr. Richard Vincent Donahue, USN, (ret)
Who saw to it that I completed this
research and wrote it all down.

FIELD EFFECT

Table of Contents

FIELD EFFECT

Field Effect

. . . The Pi Phase of Physics

by
Leigh Richmond Donahue

. . . An analysis of the structure of the electron which brings to light the anatomy of black holes and superstrings.

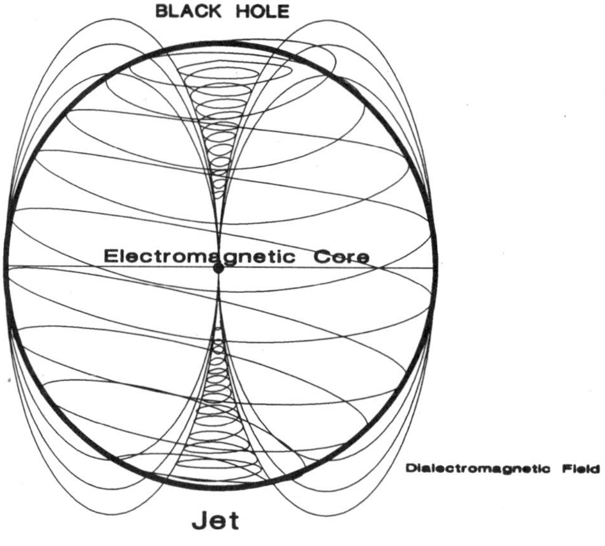

The Electron, the earth, the galaxy, and all other living things

Whether you're referring to an Angel or an Apple a Rock or a Frog; The "Field" is complete...around and through each and every one.

1
Evolution 101

People think of evolution as something that happened in the distant past, and that took millennia to be effective.

Not so.

One of the most comprehensive and abrupt evolutionary steps was taken by the human and his civilization within the past seventy years, though it is hardly recognized for what it is. Another such step is just being taken.

Mankind evolved subject to a 7.83 cycles per second pulse. That is the electronic pulse-rate of planet Earth. Mother Earth's heart-beat, if you like—and the comparison is apt.

It is a constant pulsing, very like the constant heart-beat of your own body.

Man evolved in tune with that pulsing, from the beginning. The heart-beat of his mother—and humankind in her womb.

It was not until this century that man developed his own pulsing—electromagnetic currents that began to override and blot out the pulse-beat in which he had been nurtured; until he began to develop his own civilization-wide heart beat. Until his own electronic pulse began reverberating through the individual bodies and the body of the civilization.

You have heard it said that more progress has been made in the last seventy years than in all of history combined. And that is true. But did you look—did you ask—what was the factor that initiated that dramatic change? What made it possible—what made it inevitable?

Early this century alternating current electricity was developed; and almost overnight—in planetary terms—this nation was wired for sixty cycle alternating current, from coast to coast.

You may think of electricity as within the wires in which it is carried—the cross-country wires, and the wires within the walls of your home.

It's carried in those wires, yes. But the magnetic field that pulsing electricity creates reaches far beyond those wires, and it pulses in your body, day and night, invisible and unrecognized, almost wherever you are. From birth until death you "dance" now to that sixty cycle pulse.

The sixty cycle pulse is a stimulant on the two halves of the cerebrum. A very real stimulant. A stimulant that induces the brain to higher and more effective activity—for those who can handle the stimulation.

For those who can't—well, they're left behind. Drop-outs. Has-beens in the evolutionary sequence.

It was in the late 1800s that electricity was first introduced, as a direct current. That had little observable effect, since there was no pulse involved.

Then the Tesla alternating cycle generator was developed, capable of sending AC current across the miles.

It was almost happenstance that the alternating current in this country was set at sixty cycles per second—not because it was beneficent, but because the first Tesla alternating current generator was set at that frequency.

Sixty cycles were used in this nation; a fifty cycle per second pulse rate was used in most of Europe and the rest of the world. That was less effective on the brains.

By the late 1920s, most of this country was wired for AC electricity. It pulsed at the human from the wires in his home or office or factory; through the long cross-country wires; from the generating stations that sprung up everywhere.

Pulsing in his blood; pulsing in his brain. Wherever he went, he became subject to that stimulus.

There were scattered effects from the first installations—before the '20s.

But have you noticed what happened when this whole nation became wired, and, wherever you were, you "danced" to that sixty-cycle pulse-rate, over riding the pulse of the Earth?

Almost overnight our brains came alive. We went from the horse— ten miles an hour—to the car, to the plane. From the horse-drawn plow to the tractor. From communication by ship to communication by satellite?

From the very slow lane—walking or horseback—to the **very** fast lane. In little over seven decades.

We went from the outhouse to indoor plumbing. To a network of roads to accommodate the new vehicles. To buildings that soared into the skies. To farmlands that stretched from hither to yon. From medical house calls to cat-scans, to organ replacement. From washing done by boiler and scrubboard to washing machines and dryers. From wood stoves to radar ovens. From arithmetic to computers. To the telephone, the fax machine.

It has been a very short maturization; but the human has become a new breed of cat.

Comes now the new electronics of the field. Scalar wave electronics. The electronics of the field half of the human as versus the electronics of the body.

The human is, in itself, an alternating frequency dynamic. The body, the spatial half, is only one-half of the equation known as human.

The other half is the field, with its own holographic brain, its own senses to feed that brain; its own communications potentials.

Now, developed over the last thirty years and slowly reaching both maturity and wide-spread use, are the "opposed electric" and "opposed magnetic" fields of the new field electronics.

Dialectric. Diamagnetic fields. Dialectromagnetic (DEM.) Scalar wave. EM as versus DEM.

EM/DEM. Equals and opposites. Yin and yang. Positive and negative.

And the DEM spectra pulse—not in the body, which they ignore; to which they are irrelevant—but in the field and in the field brain.

DEM has been under development in Russia for around the past 30 years. When the American Embassy in Moscow came under electronic attack— it was an early form of DEM. When Khrushev pounded his shoe on the table at the United Nations meeting, announcing that Russia had a new weapon that would destroy us unless we made peace, it was DEM, still in its early stages.

America was not far behind in developing the new electronics. Then both nations started bombarding each other with the spectra, using it as a weapon. China was the third to get into the act.

Whether any other nation has the technical abilities entailed is a question. The potential is there.

Privately, the new technical abilities have been taken up by some farmers – and proved themselves in the "bottom line." When the aura over lands that were farmed were treated positively, they showed less need for pesticides and fertilizers.

The basic of the DEM spectra is that everything that exists is a combination of the electromagnetic and the dialectromagnetic. We and our surroundings are made up entirely of EM and DEM spectra. Since the DEM is the patterning agent, and the EM is the demonstration of the pattern, treating the pattern results in the changes desired in the soil – or the changes in the human by a weaponry wielded against him; or the changes in his holographic (field) brain as he is subject to the pulses.

The DEM spectra have been broadcast across and through this planet in increasingly refined patterns and increasing quantity for thirty years or more.

They have been used as weapons. They are just as efficient when used with a positive potential.

Good or bad, DEM broadcasts are an expanding part of our daily lives.

These spectra do not effect the body; they affect the holographic brain in the body's field.

The field brain of the human is more and more pulsed by these scalar wave spectra; stimulated by them; more and more do we "dance" to their pulse.

There are those who can handle the stimulation; and there are those who cannot.

Meantime, it is quite possible that we are watching the next evolutionary step in the human equation—as participants.

Since we, the people, are being bombarded and stimulated by frequencies of which we are only newly aware; and since we are to be participants in a new evolutionary step for the human race; it behooves us to look at the structure of ourselves, our universe, and the DEM spectra.

2
The Pattern

We humans are part of a vast and quite complex galaxy. That galaxy is built and operates on logical patterns. Because they are logical, they are understandable.

Because we are intelligent, we can understand the basic make-up and the systems of operation on which we and our galaxy, in all the complexities, operate.

The pattern and the operative systems of the galaxy itself can be seen most easily in the pattern and operative systems of the electron.

Once you find the pattern and the operative system, the rest falls into place, without anomalies.

The pattern of the electron builds into the atom, and holds through the periodic table of the elements. It explains, for instance, why the atoms in a crystal, each in high, individual motion, remain in exact relationship the one to the other.

It holds through the vast complexities of the life forms, and illustrates where the difference between non-life forms (rocks, crystals, planets, etc.) and the life forms (from amoebas to humans) begins.

That pattern holds, even, through the differentials in the human – his self-awareness and intelligence.

The pattern itself is built from four elements:

1. The ether – a sea of energy without motion, which exists throughout the universe.

2. Energy in motion: What we call an electrical flow. A twisting thread of energy moving through the sea of ether.

3. Magnetic force: Ether forced into motion by the electrical flow. This takes two forms:

(a) As the electrical thread pushes aside the ether, it pulls some of the ether into motion around itself in a spiral coil of magnetic force, much like a coil of wire, stretching its length. This coil is at right angles to and moves in the opposite direction to its electrical center. As the energy of the electrical flow increases, the coil tightens. When that energy relaxes, the coil loosens. This is known as frequency variation, and can be seen in a comparison between a radar and a radio wave.

(b) Meantime, the magnetic coil causes ripples in the sea of ether around the now electro-magnetic flow. These ripples spread out around the electromagnetic path, decaying towards infinity, like ripples from a stone dropped into a pond. This the magnetic field.

4. Vacuum: Where electromagnetic flows have cleared areas in the sea of ether.

None of these four factors is particulate. In other words, they are not made up of particles. Not the sea-of-energy without motion (the ether;) nor the energy in motion (the electrical flow;) nor the energy of the ether forced into motion (the magnetic shield and its magnetic field;) nor the vacuum created in the ether by the various flows.

Granted these four components, the structure of our alternating current universe becomes understandable.

The original action and reaction of the four components follows this pattern:

A twisting thread of energy (electrical) flows through the sea of ether, pushing the ether aside, creating a magnetic coil of motion at right angles to itself that moves in the opposite direction. This magnetic coil has the characteristics of a wire coiled around and along the electrical flow.

The energy of motion has now become an electromagnetic flow. It creates ripples in the sea of ether moving away from its path in a circular motion, which are know as its magnetic field.

As this electromagnetic thread moves through the ether, the leading electrical center becomes blunted. Instead of pushing the ether aside, it builds a bow-wave which becomes a wall.

The electrical head meets that wall, flares and collapses into a flow back along the path it has just cleared through the ether, pushed on by its magnetic coil, which flares and collapses behind it.

As the magnetic coil is completing its collapse, the electrical thread is again being blunted, creates the bow-wave-wall, and flares and returns along its own path, followed and pushed on by its collapsing magnetic coil.

This process is repeated over and over.

This is the alternating current of electromagnetic energy that has been tamed to flow through the wire of your home.

Again: It is NOT particulate. It is not made up of electrons or any other particles. It is an alternating current of electromagnetic energy.

FIELD EFFECT

3

The Electron

To build a universe, you need a **stable** particle as its base.

The **pattern** of that particle must be obvious in everything from the structure of the elements to the galaxy, including the various solar systems, the rocks and the humans.

It must be of a configuration that cannot easily be destroyed. It must have a gyroscopic configuration and precess when confronted with obstacles, rather than battering or being battered. It must create itself continuously from the materials at hand – the non-particulate energy in motion (electrical flow;) the sea of energy without motion (the ether;) the energy forced into motion, the reactive flow (magnetic force;) and the vacuums created as the flows interact.

It must be in motion, thus qualifying as energetic. It must have a push and a pull of its own (polarity.) It must respond to both a push (a kick in the pants) and the pull of a vacuum, or of surrounding energies. It must have a near-infinite lifetime.

Such is the electron.

There are an infinitude of minute particles, with half-lifes in millionths and billionths of a second – quarks, charms, bits and pieces. They do not meet the criteria. They are strays.

It is the electron that is stable; and it is the pattern – the configuration of the electron that can be found through the periodic table of the elements; through the molecules; up through the animals and you; and it is the pattern of the galaxy.

The electron is constantly being created in its infinite numbers from the four materials named: the sea of ether; the energy in motion; magnetic energy forced into motion; and the vacuum caused by the various motions.

The galaxy itself is constantly being created from those same electrons. The opposed forces at Galactic Center continually throw off torrents of electrons that spiral outwards to combine into atoms, molecules, suns, planets and us.

Then just what is the structure, the pattern of action and reaction, that is the electron?

Take the twisitng thread of electrical energy when that energy meets the wall of condensed ether, (its bow wave). This time instead of turning back along its cleared path, it turns in a circle to meet its own tail. It is still surrounded by its magnetic shield at right angles to itself and moving in the opposite direction, which keeps the electric fields from combining.

This is the snake of Chinese antiquity that swallows its own tail.

This thread of the now electromagnetic energy, its center flowing in one direction, its magnetic shield flowing in the other, meets and passes its own tail. The coils of the outer magnetic "wire" of the head and tail meet and mesh like gears, pulling the electromagnetic flow into a tighter and tighter ring.

As it tightens, the ring becomes dense and denser, until it creates so strong a vacuum at its center that an amount of ether, equal to the amount of energy in the ring, is pulled in at one side, through the center, and out the other side.

With this opposed flow going through its center, the ring cannot either become smaller or collapse. This small, tight, extremely dense ring of electromagnetic threads, with an intense vacuum at its center, is the core of the forming electron.

Now we look at the opposed flow, pulled through the center of the core from the surrounding ether. As the ether goes into motion, its threads each form a magnetic shield.

This opposed flow, this dialctromagnetic flow, spirals through the center of its core, jets out spherically to a tremendous distance (in respect to the core) and is pulled back as it meets the wall of ether surrounding the entire configuration, to the vacuum at the nose of the core-ring; and through again in a raging, pico-sized torrent. Its motion creates a black hole at its nose, a jet at the tail, and a volumnious sphere of threads of magnetic-wire-shielded electrical energy around the core.

This sphere of opposed energy is the field of the now-formed particle: Its dialectromagnetic half. You have here a plethora of opposed energies in a variety of motions, interlocked into a dynamic whole that is nearly indestructible.

This is the electron.

The two halves of the electron are its electromagnetic core, a tiny, dense ring at its center; and its dialectromagnetic field, a volumnious sphere flowing through and surrounding the core.

You now have—in pico time and pico space—two energy flows inextricably interlocked. The energy flows are opposed in velocity, in volume, and in directions of flow. They are in a gyroscopic configuration, and will precess when confronted with obstacles, rather than battering or being battered.

This electron, with its high motion, creates a vacuum in the sea of ether around itself. In flowing towards that vacuum, the ether pulls electrons towards each other. This is the force known as gravity.

The field of this particle will respond to the pull of a vacuum, trailing its core. The core will respond to a push—that kick in the pants—trailing its field.

The forces will remain interlocked and stable unless they are pulled or pushed to a velocity at which the field can no longer reach the nose of the core—such as in a cyclotron or other catastrophic violence. If this happens, they will dissolve into stray energies with little half-life.

This particle has a "black hole" at its nose where the field spirals in; and a jet stream at its tail where the field spirals out. It moves through the ether propelled by its jet, at the velocity of light.

It is constantly created from the materials of the universe— from the ether, from electrical flow, from magnetic forced-flow, and from vacuum, and the "winds" of its passing stimulate ever-widening spheres of that natal electrical flow.

This is the electron. The basic **stable** particle.

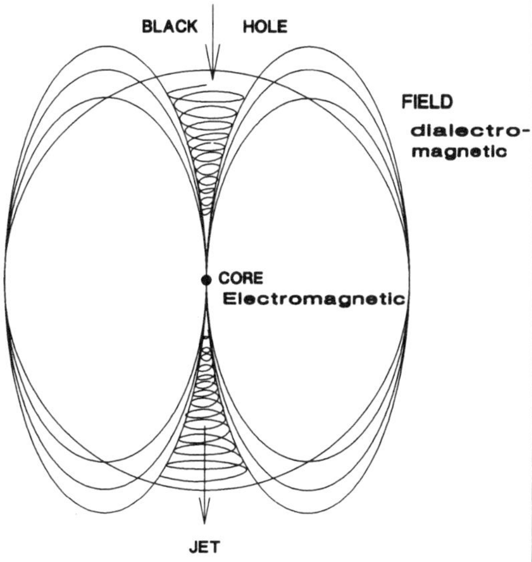

The ELECTRON - Shown in two dimensions

The electromagnetic core and the dialectromagnetic field -- equals and opposites, interlocked -- operate as a unit of opposed forces.

You have here two forces—the visible and the invisible.

The core of the electron is the spatial half of the duo. It is electro-magnetic. It is the stuff of which planets and people are made. It is small, (pico-small,) tight, and has the apparency of a solid, although it is an energy form in very high motion. Though they are interlocked and indivisible, the field is invisible to the core.

The field is the time half of the equation. Instead of being electro-magnetic, it is dialectro (opposed electro) diamagnetic (opposed magnetic,) in structure. It is voluminous, and the strings of which it is composed interlock with the field-strings of other electrons to group the particles; to capture the wayward, to create the beginnings of cohesiveness, and to form a gyroscopic barrier around the cores so that each core remains a separate structure. The core is invisible to the field.

The fields hold the cores in a past-present-future sequence, as well as in patterns of relationships the one to the other—as, for instance, in a crystal.

Two forces—one the spatial element, the other the time and patterning agent—indivisibly interlocked, in gyroscopic motion, opposed in velocity, volume and density; and invisible the one to the other.

This particle—the electron—moves through the ether with a black hole at one end and a jet stream at the other, at the velocity of light, precessing from obstacles gyroscopically, captured when its field is held by vacuum (gravity) and grouped when the strings of its field interlock with the strings of other fields.

4
The Proton

The proton is born when one electron swallows another.

Mostly when electrons meet they precess away from one another. They bang fields and jet in another direction; or they bang tails and get gone at very rapid velocities.

When the jet stream of an electron that has the same handedness – for instance, a left-hand spiral to its field, a right-hand swirl to its core – meets the black hole of an electron with the same left-handed field, right-hand core, again they throw each other the hell out of the way.

Not so when the jet stream of one electron meets the black hole of an electron of the opposite handedness.

Take the case of electrons A and B. A has a left-handed spiral to its field, a right-hand swirl to its core. B has a right-hand field spiral, a left-hand core swirl.

If the jet stream of B meets the black hole of A, B's jet and its spiraling field will "screw" into A's black hole like a bolt screwing into a nut.

That's fine, except for one thing. B's jet stream and field can screw right on through A's core; but B's core gets stuck in A's throat. It can't go through A's core. The two are swirling in opposite directions. So the two cores just sit there, caught.

Now you have a fairly complex configuration. B's field spirals on through first B's core then A's core, and outside A's field, around, and back in at the nose.

FIELD EFFECT

Proton – Interior Configuration

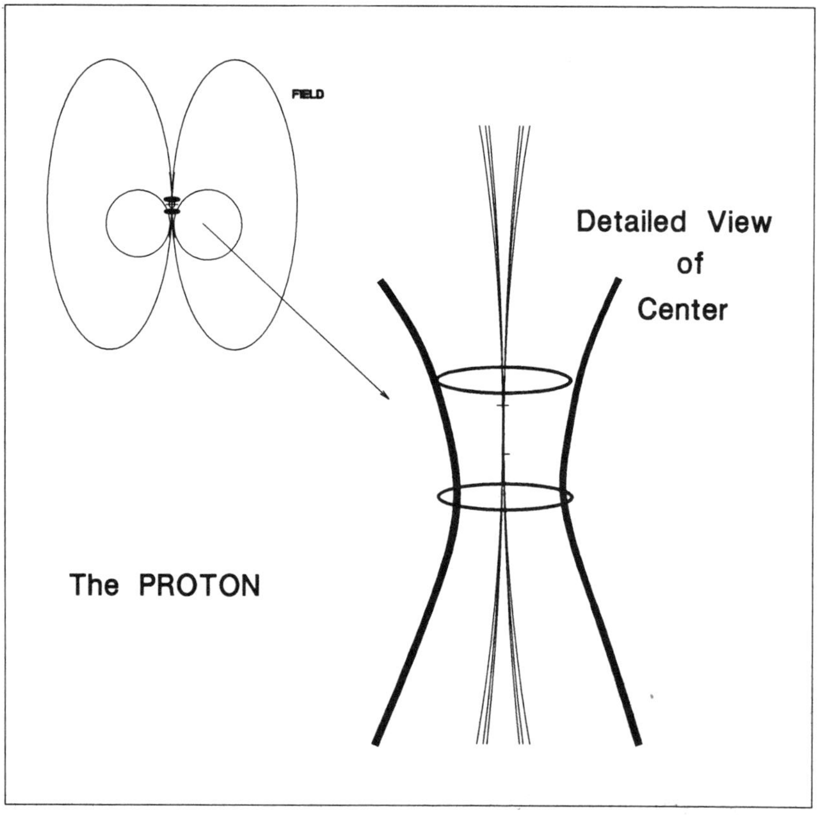

A's field, spiraling in in counter-rotation to B's field, goes around B's core and between the two cores, cushioning them apart. It continues its spiral through its own core, out and around within B's field, thus creating a shell within a shell as it spirals back into the nose of its original cone.

The energies are violent, their structures, motions and velocities opposed. The two electrons are hopelessly interlocked.

This is the proton.

What magnificent and complex furies at sub-microscopic levels and in micro time are to be found in the elements that make up our beings.

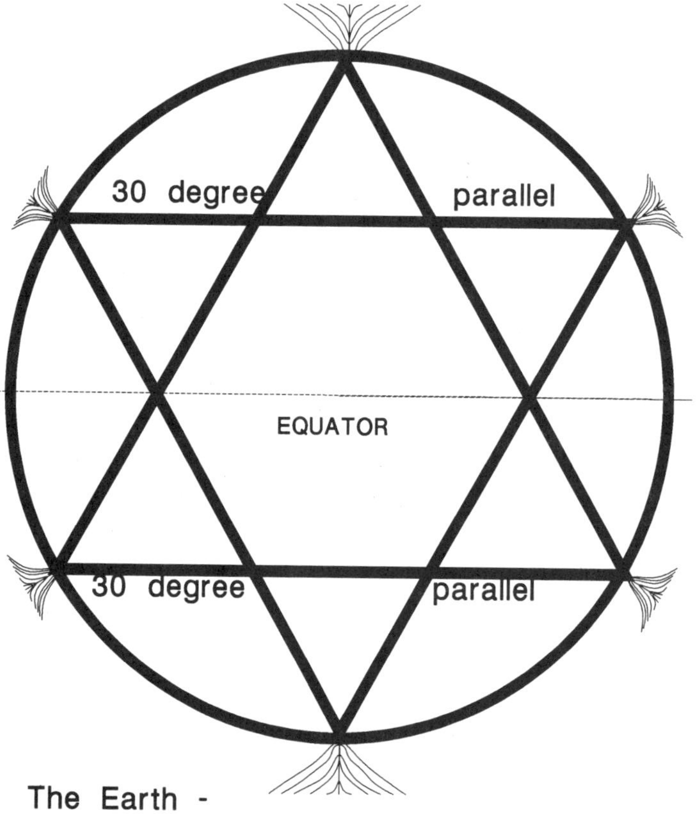

The Earth -

A complex "neutron"

The field comes in at each pole, and seeps out at the 30th degree parallel.

5

The Neutron

What happens when electron meets electron nose to nose? In other words, black hole to black hole?

If their handedness is the same—if both have either left-hand or both right-hand spiraling fields, they swirl away from each other. The winds that create the suctions of their fields tip them apart and throw them in opposite directions.

But when a left-handed electron meets a right-handed electron, nose to nose, they dance in tune.

The violence of the field entering each of the black holes cushions them apart. The suction of the black holes keeps them together. The jet of each pushes it against the other.

This is the neutron.

FIELD EFFECT

6
The Atom

It is now possible to structure the entire periodic table of the elements.

We begin with the hydrogen atom, and the proton from which it is formed—the simplest proton, with only one shell, which lies between the field of the original electron (A) and the field of the captured electron (B.)

Within this shell and the violence of its conflicting currents lie all the ingredients for the creation of electrons. It is inevitable that a new electron will be formed. The core of this new electron is dense, and will swing down to orbit the equator of the proton.

It is said that the electron has 1/1837 the mass of the hydrogen nucleus; but that its volume is 1837 times as great.

So now we must look at a definition of mass, as versus that of the field.

Mass is the dense core of the electron, and/or the core of the proton.

It is the core which is dense to near solidity.

The field(s) are, to all intents and purposes, invisible. They are so light and voluminous in nature that they are measured as a charge.

We are speaking of micro time and micro space, and their velocities.

The proton core, or nucleus, remains in the center of the atom, measurable as itself.

The electron core, orbiting the nuclcus at tremendous velocities, has the apparency of a circular volume, the volume of its entire orbit—1837 times as great as that of the central nucleus.

The hydrogen's orbiting electron swings on its field, as on wings, that stretch from pole to pole of the atom—but always within the shell to which it is confined. Those wings are of violent motion, but so voluminous and so light a density that they register only as a charge.

Thus, the hydrogen atom, the start of the ladder of the periodic table.

To begin the atomic climb up the periodic ladder, it is necessary to add energy to the hydrogen nucleus.

A proton adds energy and mass in the same manner by which it originally became a proton. It captures another electron from outside, through its black holes, into its central core.

The core of the newly captured electron is stuck in the throat of the proton, and its field forms a new, counter-rotating shell around the earlier shell(s).

With this added energy and mass, the spin of the proton increases, it becomes internally denser and smaller; but the new shell brings its volume back to the original size.

Within this new shell, again, the violence of conflicting, counter-rotating currents creates new electrons, their cores swinging in to orbit the equator of the nucleus, while their fields expand to fill the shell.

Where there are multiple electrons within a shell, they may combine to become neutrons, positrons, etc., as well as remaining electrons; but their number is restricted to the spatial characteristics within the shell to which they are confined, since their expanding fields will fill the space available, and obviate the entrance of new electrons.

Thus you have the means and structural capabilities of the ladder which we know best as the Periodic Table of the Elements. . .the atoms being built through time and evolution from the ether, through the basic stable particle, the electron, to the simple hydrogen atom— building blocks of the universe.

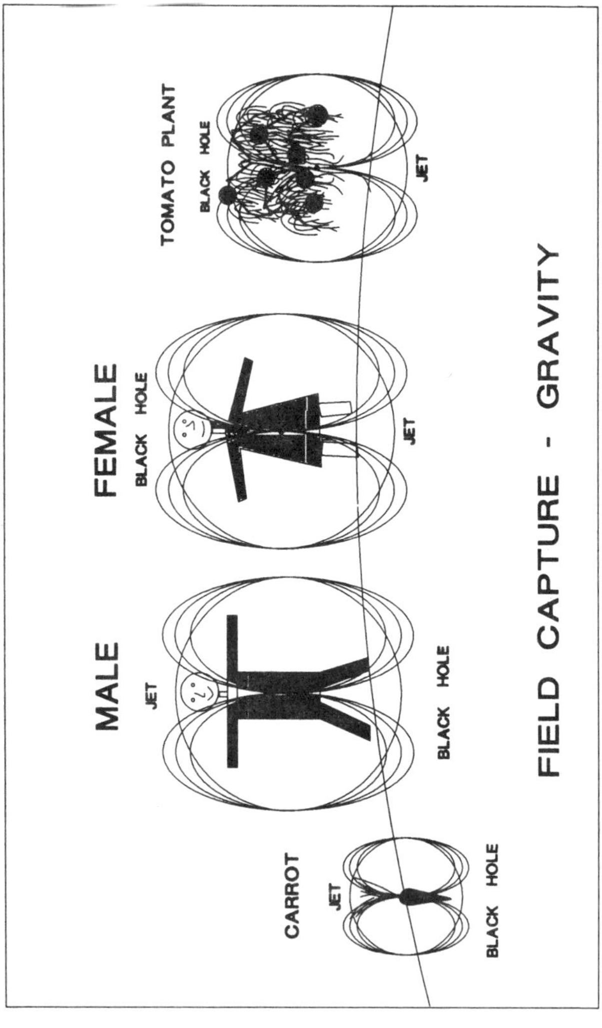
CARROT
JET
BLACK HOLE
MALE
JET
BLACK HOLE
FEMALE
BLACK HOLE
JET
TOMATO PLANT
BLACK HOLE
JET
FIELD CAPTURE - GRAVITY

7
Patterns

The pattern of the atom is the pattern of the galaxy, and of everything in between.

What you see as the galaxy is the solar and planetary groups being formed from the electrons created at the galactic center, and spiraling out through the galactic shells.

What you do not see is the network of field-strings that group the various cores, interlocking and caging them into forms of violent motion. Nor do you see the voluminous galactic shells encompassing the spiraling groups, raging in at the nose of the galaxy, raging out at its tail.

The galaxy is constantly creating itself: creating floods of elections that spiral out at tremendous velocities to form atoms, suns, planets and the myriad forms that we know.

Throughout the variety of structures of the galaxy, the same pattern is reproduced. Each atom, each molecule, each part of the intricate whole is composed of field and core.

As the atoms unite to form what eventually become suns and their planets, they move further out the spiraling galactic arms.

By the time you reach one of the far arms of the Milky Way, if you look closely, you will find our solar system and its planets on which life forms have developed.

Because we are so far out on an arm of the spiral, we know there has been time for evolution. Migration out to the arms is not a rapid process. It gives time for thc formation of those suns and planets, then the slow development through to structural validity, and then to life forms. The center of any galaxy will be found to be a chaos of beginning structures.

The pattern of development from atom to other forms is consistent: Each atom is caged in an overall network of twisting threads of dialectromagnetic energy. As it joins other atoms to form a molecule, the threads of the atoms intersect at various loci and anchor-point to each other. The suction of their joinings creates an overall field for the molecule to which the fields of the atoms become subservient. This development is consistent up the ladder of complexity from the molecule to the rock, to the planet, to the human being.

The network of the fields of the various components – say, of a crystal – anchor to each other where they intersect (superstrings) so that the atoms, each in violent motion, remain in stable configuration within the crystal. This is known as dynamic stability, as the atoms of a crystal (again, for instance) constantly rebalance themselves so that the crystalline structure is maintained.

So, too, the human being. His body is made up of atoms, molecules, etc., each with its own field, in motion, anchor-pointed to a unity with the various parts, which are in turn anchor-pointed within and subservient to his overall field. But the human has won to freedom of movement, independence from his environment. Even so, the strings of his overall field anchorpoint – though tenuously – with the fields of the people and structures around him, creating a tenuous overall network of field. A family, a community, a civilization, anchorpoints into a network with an overall field of its own.

The human has developed some ability to release the anchors although he finds a new stability in anchoring into a new environment. This may take time. Instance: Jet lag.

These are the patterns of action in the field/core configuration, and the pattern holds from the electron to the galaxy, including life forms and the human.

A pattern is **not** an identity. It is the pattern that is consistent.

FIELD EFFECT

8

The Structure of the Human

The structure of the human follows the pattern of the structure of the atom.

The brain of the physical body is a computer. This is not the seat of the intelligence; it is exactly that – a computer, although its complexity is vastly superior to the most sophisticated electronic computer today.

It has the support structure of the body, and is fed input by the five senses. It shares information throughout a vast physical network by talking, TV, libraries, newspapers, buildings, roads, art, music, etc. Information received by this computer is stored biochemically down to the atomic level, as well as in the brain itself.

This is the physical half of the human equation.

The field has its own computer. It is of a holographic nature – what you see as the aura. This holographic field computer surrounds and interpenetrates the physical brain. Its input is from the field senses.

It stores information received holographically throughout the field, and shares information through what I call the "field network" – the superstrings of dialectric/diamagnetic energy that anchorpoint it to its environment and through which it can communicate (as on a telephone, TV, fax, or computer network) with the most distant point. This network is a storehouse of information that extends through time (as the physical network extends through space.) The dimensions of the field are past, present and future, as the dimensions of the body and space are vertical, horizontal and forwards/backwards. The two—the core and the field—intersect at only one point—present time and present space.

The two computers interface—probably at the pineal gland, a crystalline structure within the physical brain that is also central to the field computer. The pineal gland shatters at the death of the body.

The intelligence, personality and character lie at this interface.

Each computer feeds its information, gathered through its own senses, to the interface. Where the two sets of information do not corroborate each other, the information is thrown back into the originating computer, where it is stored as uncorroborated. For instance, if you write a letter to a close friend and impart the information telepathically, the information will be received only in the field computer and will lie there until the letter arrives to corroborate the contents; though the friend may "know" as a hunch, that the information is there.

(This system of double input was used in the early days of computers, and possibly still is used, although of that I am not sure. When you wished to eliminate noise from stray energies from a computer input, you arranged two parallel inputs. If a stray bit of energy hit one input, it hit the second one a microsecond later, and was thrown out as noise.)

The core and the field are interlocked, down to the level of the electron. The beginning of the life forms occurs when the core and the field begin to interact, as well as interlock.

It is through a growing interaction between core and field that the increasingly higher life forms have evolved into the human.

The intimacy of interaction between body and field is an exact measure of the level of intelligence of the entity.

The business of life is the acquisition, demonstration, utilization, storage and retrieval of information.

* * *

Today we are not on information overload in either computer, though both body and field computers are fed tremendous swatches of information daily and hourly—the two computers can each handle that. It is the two support structures that are being battered to near imbecility by overcrowding.

The physical overcrowding is apparent. Not only is there the overcrowding and tensions of the streets and highways, there is overcrowding in our living styles.

In our apartments and buildings, electromagnetics signals invade—TV, radar ovens, computers. Then there are the dialectromagnetic signals (emotions and telepathy broadcast continually from the people in offices and apartments above, beside and below. It is a telepathic and emotional bedlam in the office, the apartment, the home.

People are hurting. They are bewildered from the constant and chaotic impact—and they, in turn, broadcast back the anger, hate, guilt, fear and apathy they receive across the bands.

The computers can handle the information—it is the two support structures—body and field—that are staggering.

FIELD EFFECT

9

Analysis of the Upcoming Changes

Analysis at the Centric Foundation indicates that the upcoming changes are the birth process of a new civilization.

The salient points which indicate this are these:

1. It can be shown that a civilization and the planet that bears it are each a living entity with direct analogies to the human. See Exhibit (1).

2. Recognizable in hindsight are the events of the patterns of reproduction. See Exhibit (2).

3. Planetary gestation is measured in planetary time—sequences in which 13+ years are equal to one human day. We are looking at similarities of pattern with a macroscope, rather than the microscope used in viewing human gestation; and we are looking from the point of view of the fetus, rather than that of the mother.

4. The Changes we now face have been accurately predicted for thousands of years, including the timing. One of the few events whose timing and content can be accurately predicted well in advance is a birthing, once conception has occurred.

5. Space ships from other planets—which can be shown to be intergalactic sperm—ceased landing on this planet and taking over when the ovum became fertilized, circa 1450 B.C. They continue, however, to arrive and send down their landing vehicles (UFOs) which we see in our

skies. Intergalactic sex has not ceased during pregnancy; but the sperm no longer become part of the body. See Exhibit (2).

6. This civilization has been stripping the planet of its nutrients. Now, at the ending of the gestation, the amniotic fluid (our atmosphere) has become clouded. In a human the amniotic fluid becomes clouded with pituitrin, the birth-stimulant, as gestation nears its end. The structure that has supported, nourished, and provided oxygen and blood to the fetus during gestation is being sloughed off to become the afterbirth. False labor has occurred, and true labor has begun. Predictions into the future, beyond the Changes, bear out the pattern of a birthing. See Exhibit (3).

7. The similarities of the current planetary happenings to human birth pangs and contractions, as seen from the point of view of the fetus, are striking. The late Dr. Iben Browning emphasized that current planetary upsets are not due to the ravages of man but are planetary conditions. Whether our planet is both womb and mother, or one of the wombs of something greater, we have not attempted to analyze.

The following Exhibits provide a more detailed analysis of this thesis.

Exhibit (1)

I first recognized that a civilization is a living entity in 1963, while working on the '**The Lost Millennium**'. It became obvious that a civilization spawned by sending space ships to a planet of another star. If it spawned, it was a living entity. I thought at the time that the civilization would be at about the evolutionary level of a fish.

In 1964 I was investigating the circuitry of the human body. (Computers were very young then.) I asked myself for a picture of an autonomic circuit, such as the ones I use to touch-type. Thirty-six hours later I got a movie.

The movie showed a young, blond girl, high-heeled, alert, happy, making her way jauntily into the high-rise building of Consolidated Power and Light in New York City. On her back was a sign: "Minor circuit, part of a major system."

She stepped into an elevator. The operator bore a sign "Minor circuit." She worked at her job, then went to lunch where the waitresses were labeled "Minor circuits of another system." All the people around her bore such signs. At night she bought a newspaper (No TV, those days), minor bits of the communication system, to give herself the news. She bathed, dressed and got ready to go out with a young man as part of the business of reproducing herself. The movie was much more detailed and longer.

Whoooee. The parts of the civilization were directly analogous to the workings of the human body.

A great deal of research followed, convincing me more and more of the fact that a civilization is a living entity, analogous to the human.

Then came predictions and the beginnings of the Changes. And the history of Atlantis.

Was Earth, then, a womb? Or a womb-mother all in one? If just a womb, then what was the rest of the body? That last is as yet undetermined. I shall settle for womb-and-mother until after the Changes, when we can perhaps see further.

My blond has illustrated for me relationships that fitted so well I realized it was more than just a pattern. She illustrated to near-perfection her own small interactions with other major circuits, while performing her function in her own. It was as intricate and as balanced as any body; its circuitry and functions comparable to those of the human organism. It was a tool—a macroscope—with which to look at ourselves.

Exhibit 2

Planetary Gestation in Hindsight

A civilization reproduces by sending space ships to the planets of other stars. Intergalactic sex, and the space ship is the sperm. The planet is the womb that holds the ova, infertile at first.

The space ship lands, and for several generations its own technology supports it. (Remember we are talking in planetary time sequences, in which 13+ years is equal to one human day.) When that technology breaks down there is no support structure on the planet to make possible a continuance of the space ship technology, and the people of the ship are absorbed by the natives of the planet. Those natives start out at about the condition of the neolithic, but grow in potential as "the sons of the Gods see the daughters of men that they are fair."

The Valhalla is a typical example of the first landings: It crashed somewhere in the northwest of Europe, in Sweden, Denmark or Norway. Her people even knew when their power would fail and they would be overrun by the neolithic natives. Gotteramerdung. The Valkyrie were quite real too – co-opting men from the local battle fields to defend them when their power would collapse.

Later space ship peoples built castles, known to the locals as temples, were worshiped as Gods and were served by the natives as serfs. They began small civilizations (Egypt, Crete, etc.) that were not viable and did not last. The small civilizations died and its peoples were absorbed into the planetary system, helping to build towards a support structure that could carry a true fetal civilization.

Slowly but surely the support structure grew – mining, smelting, pottery, some means of sanitation, agriculture and ranching.

The ovum (ova) were now become fertile.

The next space ship that arrives finds a support structure that can make possible the conception of a technological civilization. The ova is fertile; the space ship is the sperm; conception occurs; and the equation is closed.

Sperm continue to arrive, for intergalactic sex is not stultified. They send down landing vehicles from the ships (UFOs) but gestation is underway, the people of the vehicles do not become part of the body-civilization. They build no "temples," start no new minor civilizations.

During the first two months after conception, the fertilized ovum creates an embryo. At around two months it becomes a fetus, creating arms and legs and organs of the body-to-be.

Exhibit 3

The Outline of the History of Our Civilization

1. The Great Flood, circa 4400 B.C. – 3500 B.C. Could be equated to the first menstruation. Certainly it washed the planet clean of the carbon 14 left over from the miscarriage of the prior civilization of Atlantis. It cleaned the reproductive system.

2. Space ships land (I guesstimate at least 100) and start small though eager civilizations which prove not viable and disappear; their peoples absorbed into the native populations.

3. Circa 2700 B.C. the space ship lands that built the pyramids. This can be equated to a true breaking of the hymen, since the pyramids established a landing grid and power external to the space ship itself. This may have been a conception that proved unviable.

Circa 2200 B.C. Sodom and Gomorrah. Created the dead sea. This was possibly the discharge of the unviable pyramid-peoples conception. Certainly it destroyed the power grid.

Circa 1450 B.C. Events were violent. The fifth planet blew and became the asteroid belt. Earth went out of orbit for 50 or more years. Things were strenuous. I equate this to fertilization of the ovum, and the violent change that takes place in the womb. Also to the fact that from then on, no space ships are known to have landed to create their own nucleus successfully.

Circa 776 B.C. The communications system left behind by the last space ship blew up. Again there were major changes. I equate this to the change of the embryo into a fetus, which is a fairly violent change; and perhaps the onset of morning sickness.

By 600 B.C. ships from the Mediterranean were making their way down the coast of Africa and across to the Americas, establishing colonies and communications. The fetus was reaching nerve and blood systems to forming its organs, legs and arms and head. There is no more sign of outsiders establishing temples and being worshiped and served by serfs, though command was by nobility etc. they were of the people.

It was before and after 1 A.D. that the Messiahs—the teachers—began to appear to introduce knowledge and ethics. The hind brain was (I find) going into operation. The Christ, Buddha, Mohammed, etc.

Warfare and violence have been major factors from the beginning, as the various parts formed, tried to maintain an autonomous position, were overrun, changed, and absorbed; as the parts—from the most minute to the major parts—struggled to remain separate, but were forced into the submission of being parts of a whole.

Today, the struggle is that of the support structure being thrown off the body proper, prior to birthing, when it will become the discarded afterbirth. That support structure is still in and among the body proper, but is dying of its own inadequacies now that its function has been served. Nature is not kind, the process is painful.

Summary

The forgoing Exhibit 3, is a paper I did in November 1990 on the Upcoming Changes. I see no reason to change my prognosis now; and even find quite a bit more evidence that it is a correct interpretation of the extant facts.

First: the giant steps in technology that have occurred as the body creates its own unity and frequencies. Second, the electromagnetic frequencies developed by the body-civilization. And third, now as birth begins, the development of the dialectromagnetic spectra—the field frequencies.

The field/body are an alternating frequency structure, with the field reaching toward maturity as the body reaches toward senescence.

The time-differential in the body, too, of the growth and death of cells (us) lengthens after birth. In a growing fetus the cells grow, replicate and die at a fast rate. After birth, the process slows. By senescence it is almost stopped.

The inevitable advent of the electronic town meeting is another signal that the body is nearing birth. The electronic town meeting will activate the nerve system that unites the body to the developing cerebrum (the government) which system will override the sluggishness of the formerly isolated cerebrum.

The hind brain, the old brain, the brain that has existed from the time we were amoebas (the church) is activating the autonomic nervous system, to see to the internal housekeeping of the body as it is born and must stand on its own. Its function is to keep the body alive under any and all circumstances conscious or unconscious, and to prevent too rapid changes.

The separation of the elite and the obsolete—the true body with its functioning cerebrum, nervous system, autonomic system,— with its technologies intact and operating—the separation of that new body from the former support structure that made it possible, that will become the afterbirth—is nearing completion. The gap between the two is widening, not without struggle and pain, but very surely.

The variety of predictions—dating back to the early days of the planetary pregnancy—assume a new validity with this interpretation of the facts.

The predictions of days of darkness before the giant "quake" which is perhaps the birthing itself. . .leaving behind, dying within the planet, the afterbirth, to be discharged.

These days of darkness, accompanied by a lack of oxygen, are quite probably the trip down the birth canal. Oxygen will be nearly absent, as the body separates from the mother's, or planet's oxygen supply, and has not yet begun to breathe on its own.

These days, and the huge quakes, are predicted to be followed by a tipping of the planet—as the newborn assumes a new position in respect to its mother.

* * *

Try Occam's Razor on this analysis

* * *

Another for instance: It has been said that the brain cells do not reproduce themselves. This is true, as you will find if you look at our government. However, what has not been noticed in the human, is that specialized cells make their way to the brain to replace the cells that are disabled or dying.

* * *

Suggestion: When the new field frequencies, the dialectro-diamagnetic frequencies (scalar wave) came into use on this planet, we defended ourselves against them. We hid behind ELF broadcasts, broadcasts of our mother's planetary frequencies, the heartbeat of our mother.

However, we are nearing birth. We will have our own frequencies very soon – both electromagnetic and (weak) dialectromagnetic. We may, as babies do, hold ourselves against our mother to hear her heartbeat, but it is coming time to listen to our own; to allow the mother's heartbeat (planetary) to become a memory.

Self image of the author.

Think Widdershins!

About the Author

This book is a summary of sixty-five years of research. It is a completely new look at the human, his structure, and his relationship to his universe.

The research started when I was fourteen. Dad was an Episcopal minister, and the rectory a brown-stone front in Louisville, Kentucky. There were five daughters and four bedrooms for them, two on the top floor. My sister Pat had one, my sister Lael the other. They cared about having their own rooms. I didn't, so I shared a room with one or the other as was convenient.

Lael was away for a couple of weeks when I decided that my ethics were based mostly on what I had been told by "authority." It was time I worked out my own standards on what I knew of my own knowing.

So I locked myself into Lael's room for two weeks, except for meals, etc. My parents were very tolerant.

The first thing to look at was the question of God. I decided I didn't have enough information. If He existed, fine. If He didn't exist, mankind had so much need of Him that they'd mock Him up in such a way you couldn't look behind. So God went on the shelf for future reference.

What I **could** know of my own knowing was that man was an evolutionary creature. You could see that in the fact that an embryo developing into a fetus and being born recapitulated the entire evolutionary cycle. There were other factors, but that was the main one. QED. First fact.

Then, if evolution is a fact, devolution was necessarily also a fact. You could see that – among other things, in the great apes. They have the fully convoluted cerebrum, but without the connections between that make for intelligence. You don't get the fully convoluted cerebrum until you reach a high state of evolution. Therefore, they had necessarily reached such a high state, and then devolved. Second fact.

Next came the recognition that man has progressed, evolutionarily, to this point, but it's not a closed equation. There's more ahead.

That gave me my ethical standard; What contributed to evolution was good (I decided I was on that side of the equation.) What contributed to devolution was per se bad as far as I was concerned.

Then came another factor. You take a good mind in a good body and there is a tremendous potential. And it is hardly touched. Why?

That was the question I would tackle.

Eventually, down the years, I picked out a self-portrait that is a tree frog, ogling the world, his neck stuck out. And a motto: Think Widdershins. Think outside the normal patterns and perspectives. Take a look behind the facade of usual attitudes.

When I went to college (only for a year; the Big Depression obviated more) I was going to take an MD and become a neuropsychiatrist. There I discovered that doctors were mechanics and, as far as I could tell, psychiatrists made no sense. Not the professors; not the practicing professionals; not their children. I spent weekends at the state insane asylum, and was unimpressed with what was being done.

I was still a very new undergraduate, all of sixteen years old, and probably very impertinent. but those were my conclusions. Neuropsychiatry was not for me.

That was when I chose anthropology—the study of holistic man. (The term "man" is generic, derived from the word "human." It has no gender connotations. For gender you must go to the words —"male" and "female".)

However, it had to be more than just anthropology. The study of the holistic human included necessarily his physical makeup. Wherefore, physics.

That summer, all of seventeen, I went to New York and became the assistant manager of a women's hotel on Sutton Place, and I was there when Black Friday happened. Black Friday—when the stock market crashed and the Big Depression started. It was a phenomenon like nothing else. People jumping out of windows. Almost every hotel with suicides. We were one of the very few to escape that catastrophe. But within months the hotel went bankrupt.

The excitement was on Wall Street. They were letting people go down there by the thousands. So I went to Wall Street and got a job. With Salomon Brothers. They said they would take me on for a two weeks trial, so I asked if we could make that mutual—I would try the job for two weeks. I think they were quite startled but agreed. I stayed the two weeks, watching the stock exchange in my spare time, trying to make sense of the chaos of our system of capitalizing business. It seemed anachronistic, even then. At the end of the two weeks, Salomon Brothers offered to make me a permanent employee, but I thanked them and said they had given me a good look at finance, and I thought it enough for now.

I went back home to Brunswick, Georgia for a few months, realizing that college was not a possibility. The church couldn't even pay my father's whole salary.

The excitement was still in New York, so I went back. At that time FDR had started his first "back to work" program, and I was offered a job as assistant to a psychiatrist who was screening people for the jobs that were being created. I took a look at that and decided that that was no way to study what was happening. People were under unbelievable stress – and my place was out on the street with them. Studying. Watching. Being part of.

In those days there was no unemployment insurance; no welfare; no social security. If you were old or out of a job, out of funds, tough. You were on your own. There were a few private charities, and a few soup kitchens – if you could walk to them.

So I went out on the street, with not cent one in my pocket, all of almost-nineteen, and an anthropologist and physicist by decision (mine.)

My sister Lael was living in New York, so I told her I had a good job in Washington, but could get back once a week. This gave me a shower and a good meal once a week, and I stood by my story.

Stress. Unbelievable stress. And people were surviving – somehow.

The human is miraculous. I studied them – men, women and children – as one of them. The quirks, the make/break situations,. . .and yes, the humor. Wet, cold, hungry, building cardboard shacks along Riverside Drive, hopeless – the human and his reactions and abilities – surviving.

I'd get a small job here and there; oddities, for there were no jobs. One time I hadn't eaten in four days, and I got 75 cents. I went to a restaurant and sat there looking at the food. My stomach had been empty so long it couldn't accept it.

Do you know the worst part of being that broke? There is no place to go to the bathroom.

Finally I gave up – or rather, my sister, bless her, found out a very little of what I was doing, and yanked me into her apartment. But I had had an education like nothing else.

So I got a job on the Grace Line's Santa Lucia, her maiden voyage. I would work my way around the world. That got me papers as an able-bodied seaman. I was a waitress, but they had no other way to list us. On the West coast I would get a job on the Dollar Line, and go around the world. The Dollar Line, it turned out, used waiters.

On my second trip – and the passengers on those trips were wealthy, and tipped well at the end of the trip – we got to California just as FDR declared a bank holiday. The passengers had no money except what was in their pockets. No credit cards in those days. So, no tips.

And so it went. I married, had two children, divorced, and later married Walt Richmond, a research physicist. With him I studied physics and electronics. For thirty years, with one child. He left formal research – the paid variety – since neither of us could live under secrecy, and any basic research in those days was completely under secrecy. We decided we would starve instead, and do our own basic research.

So we starved happily for thirty years. When we got brokest, I would get a job as a waitress (you get food and money that day). I'd work long enough to get a few weeks ahead (it takes a few weeks to get your first salary on a newspaper) and then I'd get a job as a reporter. (A research physicist can't just go out and get a job as that. Walt would take care of the children, and keep on with our work.)

I'd work as a reporter until we got a few months ahead, in which to write (and get paid for) a book. Then we'd go back to our own work.

We owned only what we could put in our station wagon, along with the youngsters, and would move to wherever research in which we were interested was being done. Our books were science fiction, hard-core variety, written around our findings.

It was a fun life. People would ask us how we had the courage, and they wished they had it, to live as we did. We starved happily for thirty years. One thing we did was teach writing at a university. The physics professor there told us his students were learning more physics in our class than in his; which delighted us.

At the last, we found a houseboat. We asked the owner to let us send him $100 a week until it was paid. Then we'd come pick it up. So I got a reporting job, and – the children were grown and gone, Walt got odd jobs.

We had the houseboat, the C-Lab, two-thirds paid for when my publisher asked me what sort of contract we had. We don't I said. Horror. But what will happen if the owner reneges? Nothing on paper? "I guess we'll just have lost our houseboat," I answered.

So we lived on the C-Lab at Rathmann's Marina in Melbourne, Florida, and taught writing at the Florida Institute of Technology.

Then Walt died, abruptly. From internal injuries from an accident that we didn't know he had. Life shattered.

I kept on teaching at FIT, though I'm not good on a podium or a stage. It was Walt who was the teacher.

Come graduation time, a note came down that all the professors and adjunct professors (that was me) were to walk in the academic parade. I wrote back that I had no degree from anywhere, so I would not be able to join the others. The vice president called me, making me very proud. "You have more education than most of my professors," he said. "Please wear our post-graduate gown and hood."

A year passed. I had social security now, $94.75. My sister had written when we first began to receive that manna from heaven, that we were the only two people she knew who could live happily on $90.00 a month, and "spend the $4.75 on unheard of luxuries."

A year passed. There would of course be no more marriage for me. There'd never be another Walt.

Then down the pier came Dick Donahue, former mayor of Melbourne, Commander, U.S. Navy, retired. Just starting a political campaign. Would I help him with his brochure. I'd known him as the Mayor, but I didn't know his wife had died a few months before. He was as different from Walt as possible – and wonderful.

Six months later we were married, and he insisted I continue my research.

So here it is. We've been married thirteen years, and have packed at least twenty-five years of happiness and adventure into that time. We live now in the Great Smoky

Mountains, way up a long cove, looking down a valley to a far mountain range. Today the snow is falling.

It was near here that we met Peter Kelly, who works on the cutting edge of scalar wave electronics, and who taught me the usages of radionics.

And it is here that I have put to paper a summary of sixty-five years of research.

My self-image is still that of a tree frog, goggling at the human race; and my motto, Think Widdershins. Think outside the traveled areas, take the new perspectives.

I believe that my findings show a cohesive whole that solves the anomalies of normal anthropology and physics. I think that we can now take a look at the universe as an undivided unit, following a pattern from the bottom to the top; a rational universe. One that we have built as we evolved.

Our universe – and ourselves – make sense. Common, practical, understandable sense. Take a look.